The 2024 Supernova Viewing: An Unprecedented Celestial Event

Introduction

The year 2024 is marked by an astronomical phenomenon of monumental significance: the 2024 Supernova.

This event is anticipated to be one of the most visually spectacular and scientifically significant astronomical events of the century.

A supernova, the cataclysmic explosion of a star, releases an immense amount of energy, outshining entire galaxies and providing a rare opportunity for both amateur stargazers and professional astronomers to observe and study the life cycle of stars and the intricate workings of our universe.

What is a Supernova?

A supernova occurs when a star reaches the end of its life cycle and undergoes a catastrophic explosion.

This can happen in one of two ways: a massive star can exhaust its nuclear fuel and collapse under its own gravity, or a white dwarf star can accumulate material from a companion star until it reaches a critical mass and explodes.

In both cases, the explosion results in the star shedding its outer layers and releasing an incredible amount of light and energy.

The 2024 Supernova is expected to belong to the latter category, known as a Type Ia supernova.

The Significance of the 2024 Supernova

The 2024 Supernova is particularly significant for several reasons.

First, it offers a rare opportunity to observe a stellar explosion in real-time.

Supernovae are relatively rare events, occurring approximately once every 50 years in a galaxy the size of the Milky Way.

The last supernova visible to the naked eye was Supernova 1987A, which occurred in the Large Magellanic Cloud, a satellite galaxy of the Milky Way.

The 2024 Supernova, occurring within our galaxy, will be the first such event visible to the naked eye in over three decades.

Second, the 2024 Supernova provides an invaluable opportunity for scientific study.

Supernovae play a crucial role in the universe, contributing to the creation and distribution of heavy elements such as carbon, oxygen, and iron.

These elements are essential for the formation of planets and life as we know it.

By studying the 2024 Supernova, scientists can gain insights into the processes that govern the life cycle of stars and the chemical evolution of the universe.

Preparation for the Event

The astronomical community has been preparing for the 2024 Supernova for several years.

Advanced telescopes and observatories around the world are poised to capture detailed images and data from the explosion.

These include ground-based observatories such as the European Southern Observatory's Very Large Telescope (VLT) and space-based observatories such as the Hubble Space Telescope and the James Webb Space Telescope.

Additionally, amateur astronomers equipped with modern, high-powered telescopes will contribute to the collective observation effort.

Viewing the Supernova

For the general public, viewing the 2024 Supernova will be an unforgettable experience.

The event is expected to be visible to the naked eye in the night sky, shining as brightly as the planet Venus and possibly even rivaling the brightness of the full moon.

To maximize the viewing experience, it is recommended to find a location with minimal light pollution, such as a rural area or a designated dark sky park.

Observing the supernova with binoculars or a small telescope will reveal even more details, including the expanding shell of gas and dust created by the explosion.

Historical Context

Historically, supernovae have been observed and recorded by various cultures around the world.

One of the earliest recorded supernovae is SN 185, which was observed by Chinese astronomers in 185 AD.

Another notable supernova is SN 1006, which was observed by astronomers in China, Japan, the Middle East, and Europe in 1006 AD.

The most famous historical supernova is SN 1054, which created the Crab Nebula and was recorded by Chinese and Japanese astronomers in 1054 AD. These historical records provide valuable information about the frequency and characteristics of supernovae and highlight the long-standing human fascination with these celestial events.

Scientific Impact

The 2024 Supernova will have a profound impact on our understanding of the universe.

One of the key areas of study will be the progenitor system, or the star system that produced the supernova.

By identifying and studying the progenitor system, scientists can gain insights into the conditions that lead to a supernova explosion.

Additionally, the study of supernova remnants, or the material left behind after the explosion, will provide information about the composition and dynamics of the explosion.

Supernovae also serve as important cosmic distance markers.

Type Ia supernovae, in particular, have a consistent peak brightness, which allows astronomers to use them as "standard candles" to measure distances to galaxies.

This, in turn, helps to refine our understanding of the scale and expansion of the universe.

The 2024 Supernova will contribute to ongoing efforts to map the cosmos and understand the fundamental forces that govern its behavior.

Public Engagement and Education

The 2024 Supernova presents a unique opportunity for public engagement and education.

Astronomy clubs, science museums, and educational institutions around the world are planning events and programs to help the public learn about and observe the supernova.

These programs will include lectures, workshops, and guided observation sessions, providing people of all ages with the chance to experience the wonder of a supernova firsthand.

The event will also be widely covered in the media, with live broadcasts and online streaming of observations from major observatories.

Social media platforms will play a crucial role in sharing images and information, allowing people from all over the world to participate in the event in real-time.

This global sharing of knowledge and experience will help to foster a greater appreciation for the beauty and complexity of the universe.

Challenges and Considerations

While the 2024 Supernova offers many exciting opportunities, there are also challenges and considerations to keep in mind.

One of the primary challenges is the unpredictability of supernovae.

Although astronomers have identified potential progenitor systems and can make educated predictions about the timing of a supernova, the exact timing and characteristics of the explosion remain uncertain.

This means that observers must be prepared to act quickly and adapt to changing conditions.

Another consideration is the impact of light pollution on viewing conditions.

In many parts of the world, artificial light from cities and towns can significantly reduce the visibility of celestial objects.

Efforts to raise awareness about the importance of dark skies and to implement measures to reduce light pollution will be essential to ensuring that as many people as possible can enjoy the supernova.

Conclusion

The 2024 Supernova is a once-in-a-lifetime event that will captivate and inspire people around the world.

It represents a rare opportunity to witness the dramatic death of a star and to deepen our understanding of the universe.

Through the combined efforts of professional astronomers, amateur stargazers, and the general public, we will be able to document and learn from this extraordinary event.

As we look to the sky in anticipation of the 2024 Supernova, we are reminded of the vastness and wonder of the cosmos.

This celestial event serves as a powerful reminder of the dynamic and ever-changing nature of the universe and our place within it.

Whether viewed through the lens of a powerful telescope or with the naked eye, the 2024 Supernova promises to be a breathtaking and humbling experience that will leave a lasting impression on all who witness it.

Please use the next few pages for your notes and debates.

www.ingramcontent.com/pod-product-compliance
Lightning Source LLC
Chambersburg PA
CBHW072049230526
45479CB00009B/332